Mohamed MEZIANE

Study of Thermo-Gravitational Separation in Porous Media

Mohamed MEZIANE

Study of Thermo-Gravitational Separation in Porous Media

Interactions between Convection and Thermodiffusion and their Applications in Fluid Engineering

Imprint
Any brand names and product names mentioned in this book are subject to trademark, brand or patent protection and are trademarks or registered trademarks of their respective holders. The use of brand names, product names, common names, trade names, product descriptions etc. even without a particular marking in this work is in no way to be construed to mean that such names may be regarded as unrestricted in respect of trademark and brand protection legislation and could thus be used by anyone.

Cover image: www.ingimage.com

This book is a translation from the original published under ISBN 978-613-9-51636-0.

Publisher:
Sciencia Scripts
is a trademark of
Dodo Books Indian Ocean Ltd. and OmniScriptum S.R.L publishing group

120 High Road, East Finchley, London, N2 9ED, United Kingdom
Str. Armeneasca 28/1, office 1, Chisinau MD-2012, Republic of Moldova, Europe
Printed at: see last page
ISBN: 978-620-8-28109-0

Advanced Study of Thermo-Gravitational Separation in Porous Media: Interactions between Convection and Thermodiffusion and their Applications in Fluid Engineering

Mohamed MEZIANE

Table of contents

General introduction

Thermogravitational separation is a complex physical phenomenon that combines the combined effects of thermodiffusion and natural convection in a fluid. This process is based on the interaction between a thermal gradient and the gravitational field, which generate fluid motion and concentration gradients in multicomponent mixtures. The study of this phenomenon is of fundamental importance in many areas of science and engineering, including energy, industrial processes and the environment, where efficient separation of mixture components is required.

Thermodiffusion, also known as the Soret effect, describes the tendency of components in a mixture to separate under the effect of a temperature gradient. This phenomenon results from the differential migration of molecules according to their mass or chemical nature, leading to the accumulation of one species near the hot wall and the other near the cold wall. This mechanism was first discovered by Ludwig in 1856 and further developed by Soret in 1880, but it was not until much later that it found applications in industrial and scientific systems.

In the presence of a gravitational field, natural convection also comes into play. This is induced by density variations resulting from temperature differences within the fluid. When a thermal gradient is applied, density variations cause lighter fluid to move upwards along hot walls and denser fluid to move downwards along cold walls, creating an overall convective flow. In the case of binary mixtures, this phenomenon is more complex, as convection affects both temperature and concentration, making the flow sensitive to variations in both properties.

The combination of these two phenomena - thermodiffusion and convection - leads to what is known as **thermogravitation**, a synergistic process that amplifies the separation of components in a binary fluid. In an optimal configuration, this coupling between convection and thermodiffusion enables significantly higher separation levels to be achieved than with thermodiffusion alone. The phenomenon of thermogravitation was first brought to light by Clusius and Dickel in 1938 for gases, and has since found numerous applications in isotope separation, the petroleum industry and geochemistry, notably for the study of hydrocarbon reservoirs and magmatic processes.

In this study, we focus on the application of thermogravitation in a **porous medium** saturated with a binary mixture. Porous media offer an ideal setting for intensifying the coupling between convection and thermodiffusion due to their permeability and complex structure. The use of porous media for thermogravitational separation was first proposed by Sullivan in 1957 and Lorenz in 1959. These works demonstrated that inserting a porous structure into a thermodiffusion column could improve separation by increasing concentration gradients and modulating convective flows.

The mathematical model used in this work is based on Darcy's equations to describe flow in porous media. This model, which simplifies the classical hydrodynamic equations, enables complex configurations to be dealt with both analytically and numerically. In particular, the effect of the permeability of the porous medium, the thermodiffusion coefficient (Soret effect), and the thermal Rayleigh number is taken into account to optimize separation. Numerical simulations based on the finite element method are also carried out to validate analytical results and explore regimes where analytical solutions are difficult to obtain.

The aim of this work is to propose a new thermogravitational separation configuration using a horizontal cavity filled with a porous medium, saturated with a binary fluid, and subjected to a vertical heat flow. This device offers several advantages over conventional configurations, such as differentially heated vertical cavities: it enables more efficient separation, with better control of experimental conditions and easier cell geometries.

In conclusion, this study contributes to a better understanding of thermogravitational separation and proposes optimization methods in porous media. It also paves the way for industrial applications in the field of mixture separation, as well as in Earth sciences for understanding geophysical processes involving multicomponent mixtures under the effect of thermal gradients.

Key words: Convection; Thermodiffusion; Thermogravitation; Soret effect; Binary mixtures

Parts list

a	Diffusivité thermique ($m^2.s^{-1}$).
b	Le gradient horizontal de température ($°K.m^{-1}$).
K	Perméabilité du milieu poreux (m^2).
c	Fraction massique (sans unité).
D	Coefficient de diffusion massique ($m^2.s^{-1}$).
D_T	Le coefficient de thermodiffusion ($m^2.s^{-1}.°K^{-1}$).
m	Le gradient de fraction massique (m^{-1}).
P	La pression ($N.m^{-2}$).
T	La température ($°K$).
U, u, W..	Les composantes cartésiennes de la vitesse ($m.s^{-1}$).
$\vec{V}$	Le vecteur vitesse ($m.s^{-1}$).
β_T	Le coefficient d'expansion thermique du mélange ($°K^{-1}$).
β_c	Le coefficient d'expansion massique du mélange (sans unité).
ΔT	Ecart de température ($°K$).
μ	La viscosité dynamique du mélange ($Kg.m^{-1}.s^{-1}$).
v	La viscosité cinématique du mélange ($m^2.s^{-1}$).
ρ	La masse volumique du mélange ($Kg.m^{-3}$)

Chapter 1

Thermogravitational Diffusion Phenomena: Theory, History and Applications in Porous Media

1. Introduction :

A transport property that has been the subject of much work over the last decade is thermodiffusion, also known as thermal diffusion or the Ludwig-Soret effect [17]. This effect describes the coupling between a temperature gradient and a resulting mass flow in a multicomponent system. In a binary liquid mixture with non-uniform concentration and temperature, the mass flux $\vec{J}_m$ of component 1 contains both contributions from concentration and temperature gradient [7]:

$$\vec{J}_m = -\rho D \vec{\nabla} C - \rho C (1 - C) D_T \vec{\nabla} T$$

D denotes the mass diffusion coefficient, D_T the thermodiffusion coefficient, ρ the density, and C the mass fraction of component 1.

So thermogravitational diffusion is the combination of two phenomena [22] :

The first is natural **convection**, resulting from the application of a temperature gradient to any fluid in the field of gravity. This well-known phenomenon has been studied extensively in the past. As a reminder, the temperature gradient induces density variations which, under the action of the gravitational field, tend to move the fluid upwards along a hot wall, and downwards along a cold wall (figure 1). The result is an overall movement of the fluid in the area under consideration.

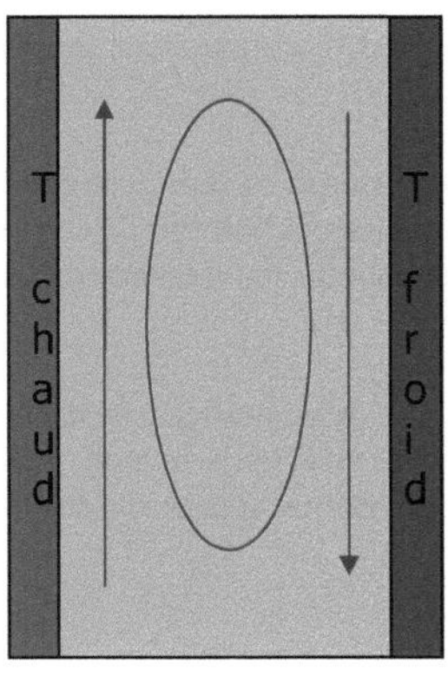

Figure 1: Natural convection in a differentially heated cavity.

We'll see later that in the case of mixtures, since convection results from density variations that are not only a function of temperature but also of concentration, the overall movement of the fluid can be more complex.

The second phenomenon is thermal diffusion, also known as **thermodiffusion**, a relatively recent discovery dating back to the last century. In the case of a multicomponent fluid, the presence of a temperature gradient generates a concentration gradient, resulting in a flow of matter within the mixture (Figure 2).

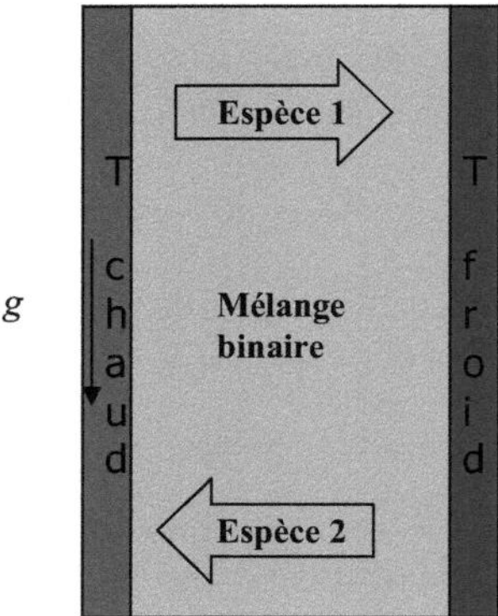

Figure 2: Thermal diffusion or thermodiffusion

In the case of a two-component mixture, we observe an accumulation of one component near the cold wall and the other near the hot wall. This phenomenon is also known as the "Soret effect".

This phenomenon leads to a separation of the constituents. This separation can be as high as 25% in stationary gas phases, when the flow of matter induced by molecular diffusion (Fick's Law) compensates for the thermodiffusion flow. Separation in liquid phases, however, is much lower.

For a long time, thermodiffusion was studied on its own, trying to avoid parasitic convection, either by using horizontal cells heated from above, or in microgravity, both to identify its main characteristics and to determine the effective values of the Soret coefficient.

In the general case, when a mixture is subjected to a temperature gradient in the gravity field, convection and thermodiffusion phenomena are present. Under these conditions, the permanent convection-related recycling of thermodiffusion-separated molecules between the hot ascending and cold descending fluid layers considerably amplifies the elementary separation produced by thermodiffusion, leading to the enrichment of one of the mixture's constituents at the bottom of the column and its depletion at the top (Figure 3).

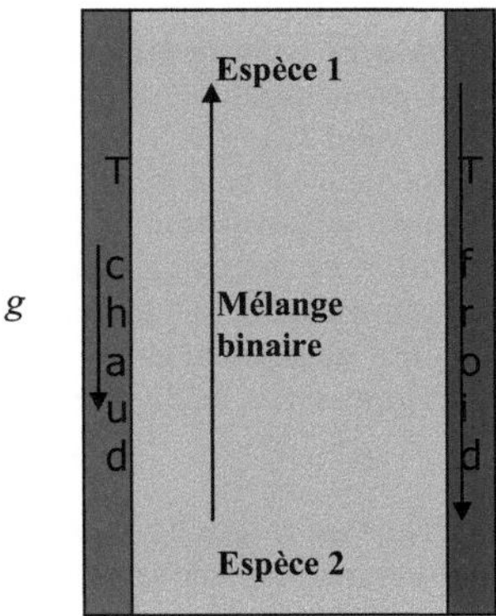

Figure 3: Thermogravitational process
Distribution of steady-state concentrations.

The amplitude of the thermodiffusion flux depends solely on the physico-chemical characteristics of the constituents and the temperature difference applied. The convection speed also depends on the geometric characteristics of the cells used. This speed must be adjusted to achieve the best possible separation. If it is too low, the horizontal concentration gradient is not disturbed, and the stationary state reached corresponds overall to a vertically stratified concentration field similar to that obtained by pure thermodiffusion. If the velocity is too high, the mixture homogenizes more rapidly than it differentiates, and any concentration contrasts that might be established are annihilated. Between these two extremes, there is a range where coupling is optimal and allows maximum separation.

2. History

In this section, we present the history of the involvement of thermogravitational diffusion in scientific research, from its discovery to the present day. The presentation is intended to be chronological and relatively succinct, and cannot therefore refer to all the authors involved in the study of this phenomenon. For further details, please refer to the bibliographies of, among others, De Groot [8] for work prior to 1945,

Platten and Legros [25] for work on thermodiffusion in free media, and Costesèque [4] for work in porous media and in relation to geology.

2.1. Thermodiffusion :

We owe the discovery of thermodiffusion to Ludwig, who first reported in 1856 [20] that he had observed a difference in concentration in an aqueous sodium sulfate solution subjected to a thermal gradient. However, it wasn't until Soret's work on electrolytic solutions in 1880 [30] and 1881 [31] that a more detailed explanation of the phenomenon, which has since been named after him, became available. He demonstrated the existence of an ionic concentration gradient in the opposite direction to that of the temperature gradient, corresponding to a migration of ions towards the colder parts of the container. Meanwhile, in 1872, Dufour discovered the effect that now bears his name, namely the establishment of a temperature gradient as a consequence of a concentration gradient, a negligible phenomenon outside gaseous phases.

Between 1887 and 1892, various geological studies took thermodiffusion into account to explain magmatic differentiation phenomena caused by thermal gradients within magmas. The results of this work were called into question as early as 1893, when the phenomenon of thermogravitation was still unknown.

It wasn't until the 1911s that thermodiffusion work was resumed and agreement was found between experimental results and those based on the kinetic theory of gases. Work on the partial separation of isotopes came much later.

2.2. Thermogravitational diffusion

It wasn't until 1938 that the general attention of physicists was once again drawn to thermal diffusion, when Clusius and Dickel [3] developed a device that used the gravitational field to generate a convection current, thereby enhancing the separations produced by thermodiffusion in a gas mixture. This method, subsequently called thermogravitational diffusion, proved to be the most effective for completely separating isotopes from certain gases. The same method was then used for liquids by Korsching and Wirtz from 1939 [18]. However, this discovery seems to have gone completely unnoticed by geochemists.

2.3 Phenomenological theory

The first phenomenological descriptions of the thermogravitational effect, based on Onsager's reciprocity laws established in 1931 [23], were given by Furry, Jones and Onsager in 1939 [16] for gaseous mixtures, and by De Groot in 1942 [9] for liquids.

This approach enabled these authors, and many others such as Bierlein [1] and Majumdar [21], to find analytical solutions for describing physical phenomena.

The development of a unifying theory of irreversible phenomena, based on thermodynamics and encompassing all diffusional processes, by De Groot in 1945 [8], Casimir in 1954 [2], De Groot and Mazur in 1961 [9] and Prigogine in 1961 [26], enabled the existence of the Soret effect to be correctly explained.

2.4. Porous media

Following the discovery of the thermogravitational effect, numerous modifications were made to Clusius and Dickel's original apparatus to improve the separation of constituents by thermodiffusion-convection coupling, and in particular to achieve better coupling between thermal diffusion and convection velocities. Of the various types of column, baffled, inclined or spiral, those with porous packing introduced by Sullivan in 1957 [32] and Lorenz and Emery in 1959 [19] are the most promising.

In 1963 [14], Emery and Lorenz developed an analytical solution to the problem of thermogravitational diffusion in porous media, along the same lines as Furry, Jones and Onsager. It enables us to predict the amplitude and kinetics of concentration variations that may result from optimal separation conditions, as a function of the medium's permeability.

These results rekindled geochemists' interest in thermogravitational diffusion as a phenomenon likely to provide new explanations for the natural migration and separation of elements. This process would probably have gone unnoticed, in terms of its possible applications and consequences in the earth sciences, were it not for the studies by Estebe in 1970 [15], Schott in 1973 [29], Sanchez in 1975 [28] and Costesèque in 1982 [4]. These authors are responsible for a large body of experimental work on thermogravitational diffusion in porous media. In particular, they proposed a method for determining the Soret coefficient, by comparing analytical results with experimental measurements on a given mixture.

This then led some scientists to investigate the possible importance of thermodiffusion and thermogravitational diffusion phenomena in hydrocarbons, as partial segregation of hydrocarbons is observed in many reservoir rocks, and the maturation of a hydrocarbon deposit is frequently associated with the existence of a significant thermal gradient. El Maâtaoui in 1986 [13], Costesèque et al. in 1987 [5] and Rivière in 1991 [27] report on experimental work on thermogravitational columns with porous packing, clearly showing that this process can be used to create significant heterogeneities in crude oil by separating hydrocarbons.

In 1989, Ecenarro et al [10] solved the equations of the thermogravitation problem in a free medium, taking into account the dependence of density on concentration in the momentum conservation equation. The effect of this dependence on separation is known as the "forgotten effect", and was not taken into account in the theory formulated by Furry and Jones. They showed that this effect can be significant in certain cases.

Platten et al. in 2003 [24] demonstrated that tilting a thermogravitation column by an angle θ to the vertical increased molecular separation. A very simple reformulation of the Furry-Jones-Onsager-Majumdar theory, in which they replaced the acceleration of gravity g by $g\cos\theta$ seems to reproduce the experimental results. The use of longer columns (1 or 2 m), higher temperature differences, and near-optimal tilt angles could lead to separations of the order of 1, even if the time taken to establish the steady state increases significantly.

Thermogravitation modeling requires knowledge of the Soret coefficient. Several authors have proposed methods for measuring the Soret coefficient.

Costesèque et al [6] set out to verify whether the value of the soret coefficient changes when moving from a free medium to a porous medium. To this end, for a given liquid mixture, they experimentally measured the time-dependent evolution of the vertical concentration gradient in the same thermodiffusion cell, once with the liquid alone and again with a porous medium saturated with the same liquid mixture. They found that the Soret coefficient was virtually the same in both experiments.

Elhajjar et al [12], on the other hand, used a horizontal cavity with thermal gradients imposed on the horizontal walls, to obtain two control parameters for better separation control.

More recently, Elhajjar, Charrier Mojtabi and Mojtab [11] have shown that, contrary to what has been done until now, it is possible to separate the species of a binary mixture in a horizontal cell heated from below with the separation factor $\psi > 0$ or from above with $\psi < 0$, and also to separate the species in a horizontal cell, without fearing the mixing of the constituents of the mixture observed in a vertical cell for fluids with a negative Soret coefficient. Finally, they succeeded in improving the efficiency of thermogravitational separation, enabling the development of an industrial separation process.

2.5. Conclusion:

Thermogravitational diffusion, a combination of thermodiffusion and natural convection, is a fundamental phenomenon in the study of fluid mixtures under temperature gradients and gravitational fields. This chapter traces the history of this phenomenon from its earliest discoveries, with references to the pioneering work of Ludwig, Soret and, later, Onsager and Furry. The evolution in our understanding of thermodiffusion and its integration into porous media has marked a major advance in methods for separating fluid mixtures.

The application of thermogravitation in porous media has opened the way to more efficient separation processes, particularly for complex mixtures such as gases and multicomponent liquids. By understanding the fundamental mechanisms underlying these phenomena, in particular the influence of temperature and concentration gradients on convective flows, it becomes possible to exploit these properties to optimize species separation in industrial and geochemical systems.

In the following chapters, we take a closer look at thermogravitation in specific configurations, introducing analytical and numerical methods for solving the equations governing this phenomenon in cavities filled with porous media. The aim is to identify the optimum conditions for maximizing separation, while considering the practical and theoretical implications of these results.

Chapter 2

Thermogravitation in a vertical cell

1.   Problem presentation :

Thermogravitation results from the interaction between a thermal gradient and a gravitational field, combining natural convection and thermodiffusion (Soret effect) in multiconstituent mixtures. This phenomenon plays a fundamental role in the separation of the constituents of a binary fluid mixture. In this study, we focus on the configuration of a differentially heated vertical cell, a reference model for the theoretical and experimental study of these processes. The aim of this chapter is to analyze this system in depth and provide an analytical solution for determining optimum separation conditions.

Figure 1 shows the geometric configuration of the problem. It consists of a vertical, parallelepiped-shaped cavity with two differentially heated vertical walls. The left wall is maintained at a high temperature T_h while the right wall is at a lower temperature T_c . The horizontal walls are thermally insulated, preventing any heat flow. Under these thermal conditions, convective flow is induced in the fluid due to the density differences associated with the temperature gradient.

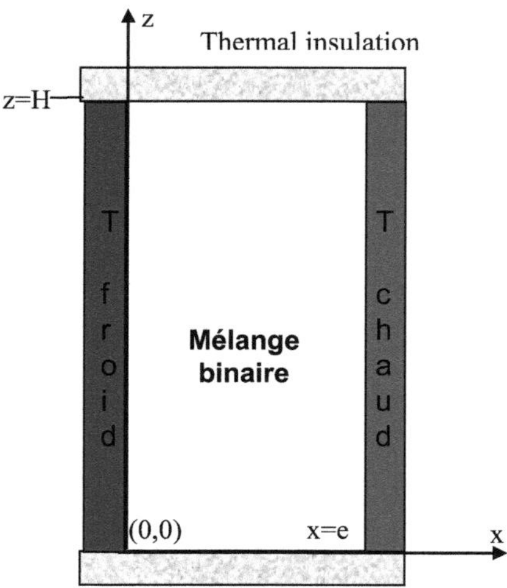

Figure 1: Vertical thermogravitation cell

The temperature field thus created generates fluid movement upwards along the hot wall and downwards along the cold wall. This natural convection phenomenon is reinforced by the effect of thermodiffusion, which causes the species making up the mixture to migrate. The Soret effect thus generates a flow of matter due to the thermal gradient, leading to an accumulation of the densest species near the cold wall, and of the lightest species near the hot wall, as illustrated in figure 1.

The system's governing equations are derived from the laws of conservation of mass, momentum and energy. These equations take into account the coupling between convective motion and concentration gradients via the Boussinesq approximation, which simplifies the treatment of density variations in the Navier-Stokes equations. The model also includes a modified diffusion equation to account for mass transfer between species in the mixture in response to the thermal gradient. This coupling between convection and thermodiffusion makes the equations non-linear and interconnected, requiring a specific analytical approach to solve them.

The main assumptions in this study are as follows:

1. Steady state: the study focuses on the steady state once the temperature and concentration fields have stabilized.
2. Unidirectional vertical flow: fluid motion is assumed to be mainly in the vertical direction, which simplifies the Navier-Stokes equations.
3. Boussinesq approximation: this approximation is used to model density variations as a function of temperature, while neglecting the effects of concentration in buoyancy terms.
4. Equilibrium diffusion: concentration gradients are small, and concentration-related density variation is neglected, which is a classic assumption in mixtures where thermodiffusion predominates.

The boundary conditions imposed on the system are as follows: constant temperatures are applied to the vertical walls, while the horizontal walls are thermally insulated. There is no material flow through the walls, and non-slip conditions are imposed, i.e. zero fluid velocity at the wall surface.

The aim of this study is to determine the optimum conditions for maximum component separation. This includes finding the optimum cell thickness to maximize separation and identifying the critical thermal Rayleigh number, which governs the stability of convective flow. One of the expected results is that separation is a function of the balance between convection and thermodiffusion phenomena, and that optimal coupling enables significant separations to be achieved under realistic experimental conditions.

However, the vertical configuration has certain limitations. Optimum cell thicknesses are often very small, on the order of a few tenths of a millimetre, making practical implementation difficult. In addition, the relaxation time to reach the stationary state can be extremely long, limiting the applicability of this method on a large scale. These limitations will be discussed in more detail in the following sections.

In the following chapters, we explore alternative configurations, such as horizontal cavities, which may offer practical advantages in terms of control of experimental conditions and more efficient separation of constituents.

2.   **Analytical resolution :**

To find an analytical solution to the problem, Furry, Jones and Onsager used a number of assumptions. In the following, we'll use the same assumptions, but with a slightly different approach.

The system of equations governing the thermogravitational diffusion process, derived from the balances of matter, momentum, energy and solute mass conservation in the Boussinesq approximation, is written :

$$\nabla.\vec{V} = 0 \tag{1}$$

$$\rho_0[\frac{\partial \vec{V}}{\partial t} + (\vec{V}.\vec{\nabla})\vec{V}] = -\vec{\nabla}P + \rho_0[1 - \beta_T(T - T_0) - \beta_c(C - C_0)](-g).\vec{e}_z + \mu\nabla^2\vec{V} \tag{2}$$

$$\frac{\partial T}{\partial t} + \vec{V}.\vec{\nabla}T = a\nabla^2T \tag{3}$$

$$\frac{\partial C}{\partial t} + \vec{V}.\vec{\nabla}C = D\nabla^2C + D_T C(1-C)\nabla^2T \tag{4}$$

In equations (1-4), $\vec{V}$ represents the velocity, T the temperature, C the mass fraction of the heaviest species, P the pressure, ρ_0 the density at the initial temperature T_0 and for the initial mass fraction C_0 , β_T and β_C the coefficients of thermal and mass expansion, μ the dynamic viscosity, a the thermal diffusivity, D the mass diffusion coefficient, D_T the thermodiffusion coefficient and $\vec{e}_z$ the unit vector along the z axis.

The main assumptions used are as follows:

- We are interested in the stationary state.
- The flow velocity is assumed to have a single vertical component depending on the single abscissa x : $\vec{V} = W(x)\vec{e}_z$ The change in flow structure at the edges of the cell is ignored.
- Assuming little change in concentration, we replace the product of the concentrations $C(1-C)$ by the product of the initial state concentrations $C_0(1-C_0)$ in equation (4).
- The temperature within the mixture depends only on the abscissa x.

- The influence of the variation in concentration on the variation in density of the mixture is neglected, i.e. $\beta_c = 0$, the so-called forgotten effect.

The associated boundary conditions are :

- Temperatures imposed on the two vertical walls :

$$T(x = 0, z) = T_F \tag{5}$$

$$T(x = e, z) = T_C \tag{6}$$

Where 'e' represents the thickness of the cavity.

- Horizontal walls are thermally insulated:

$$\left. \frac{\partial T}{\partial z} \right|_{z=0} = \left. \frac{\partial T}{\partial z} \right|_{z=H} = 0 \tag{7}$$

- All walls are solid, so there is no sliding on the walls:

$\vec{V} = 0$ on all walls.

- All walls are impervious to the material:

$(D\vec{\nabla}C + D_T C_0 (1 - C_0)\vec{\nabla}T).\vec{n} = 0$ on all walls.

Where $\vec{n}$ represents the outgoing normal on the corresponding wall.

With the assumptions already mentioned, and eliminating pressure in the system of equations (1-4), we obtain :

$$\begin{cases} v\dfrac{d^3 W}{dx^3} + \beta_T g \dfrac{dT}{dx} = 0 & (8) \\[2em] \dfrac{d^2 T}{dx^2} = 0 & (9) \\[2em] W\dfrac{\partial C}{\partial z} = D(\dfrac{\partial^2 C}{\partial x^2} + \dfrac{\partial^2 C}{\partial z^2}) + D_T C_0 (1 - C_0)\dfrac{d^2 T}{dx^2} & (10) \end{cases}$$

Where $v = \mu / \rho_0$ represents the kinematic viscosity of the mixture.

The boundary conditions become :

$$W(x = 0) = W(x = e) = 0 \tag{11}$$

$$T(x = 0) = T_F \tag{12}$$

$$T(x = e) = T_C \tag{13}$$

$$\left.\frac{\partial C}{\partial x}\right|_{x=0,e} = -C_0(1-C_0)\frac{D_T}{D}\left.\frac{dT}{dx}\right|_{x=0,e} \tag{14}$$

To solve the problem analytically, we assume that $\frac{\partial C}{\partial x}$ is independent of z and $\frac{\partial C}{\partial z}$ is independent of x.

It is further assumed that the concentration C varies linearly with z, hence: . $C = C_1(x) + \delta z$

Finally, we obtain :

$$v\frac{d^3W}{dx^3} + \beta_T g\frac{dT}{dx} = 0 \tag{15}$$

$$\frac{d^2T}{dx^2} = 0 \tag{16}$$

$$W\delta = D\frac{d^2C_1}{dx^2} \tag{17}$$

Equation (16) with temperature boundary conditions gives the temperature field :

$$T(x) = T_F + \frac{T_C - T_F}{e}x \tag{18}$$

By replacing the field obtained in equation (15), we derive the differential equation in W(x) :

$$v\frac{d^3W}{dx^3} + \beta_T g\frac{T_C - T_F}{e} = 0 \tag{19}$$

This is a third-order differential equation, and for the moment we only have two boundary conditions for the velocity, which correspond to non-slip at the walls, so we need a third condition. We use the conservation of volume flow through any horizontal cross-section of the cell, which translates into :

$$\int\limits_{0}^{e} W(x).dx = 0 \tag{20}$$

So solving equation (19) with the three velocity boundary conditions gives :

$$W(x) = \frac{-\beta_T g \Delta T}{6ve} x(x-e)(x-\frac{e}{2}) \tag{21}$$

By replacing the velocity field obtained in equation (17) we find :

$$\frac{-\beta_T g \Delta T}{6ve} x(x-e)(x-\frac{e}{2})\delta = D\frac{d^2 C_1}{dx^2} \tag{22}$$

Integrating this differential equation (22), we obtain :

$$C_1(x) = -\frac{\beta_T g \Delta T \delta}{12veD}(\frac{x^5}{10} - \frac{ex^4}{4} + \frac{e^2 x^3}{6}) + k_1 x + k_2 \tag{23}$$

Where k_1 and k_2 are arbitrary constants.

We still have three undetermined constants k_1 , k_2 and δ , and we only have two conditions for non-penetration on vertical walls, so we need to find other conditions :

When the stationary regime is established, the concentration profile stabilizes and there is no longer any net mass flow through any horizontal section:

$$\int\limits_{0}^{e} (W(x).C - D\frac{\partial C}{\partial z}).dx = 0 \tag{24}$$

We still have the condition of conservation of the mass of the solute in the cavity:

$$\int\limits_{0}^{e} dx \int\limits_{0}^{H} Cdz = C_0 eH \tag{25}$$

We therefore have four conditions (the two conditions of non-penetration at the walls and conditions (24) and (25)), and we have three constants to determine k_1 , k_2 and δ .

Condition (24) is obtained using the wall non-penetration conditions. To solve this problem, we used condition (24) instead of one of the two non-penetration conditions.

Using these conditions we obtain :

$$\delta = \frac{504\beta_T \, g\Delta T^2 v e^2 D_T C_0(-1+C_0)}{\beta_T^2 g^2 \Delta T^2 e^6 + 362880 D^2 v^2} \tag{26}$$

We introduce the separation ΔC which is the difference in concentration between the bottom and the top of the cell:

$$\Delta C = C_{bas} - C_{haut} = -H\delta = \frac{504\beta_T \, g\Delta T^2 v e^2 D_T C_0(1-C_0)}{\beta_T^2 g^2 \Delta T^2 e^6 + 362880 D^2 v^2} H \tag{27}$$

3. Optimum separation conditions :

We're now looking for the thickness that gives maximum separation.

Note that ΔC tends towards zero when e tends towards zero or infinity, so there is a thickness value that gives maximum separation.

The optimum thickness is easily found to be :

$$e_{opt} = 2(2835)^{1/6}\left(\frac{D^2 v^2}{\beta_T^2 g^2 \Delta T^2}\right)^{1/6} \tag{28}$$

And the maximum separation is given by :

$$\Delta C = \frac{(2835)^{1/3}(D^2 v^2 \beta_T^4 g^4 \Delta T^4)^{1/3} D_T C_0(1-C_0)}{270\,\beta_T g v D^2} H \tag{29}$$

 To illustrate this result, we will consider a binary mixture already used by Platten et al [24], a water-ethanol mixture with 60.88% water by mass. The mean temperature of the mixture is 22.5°C, and at this temperature the properties of the mixture are given in Table 1.

ρ	β_T	β_c	D	D_T	v	a
935.17 kg.m^{-3}	7.86 10^{-4} °K^{-1}	-0.212	4.32 10^{-10} m^2.s^{-1}	1.37 10^{-12} m^2.s^{-1} . °K^{-1}	2.716 10^{-6} m^2.s^{-1}	10^{-7} m^2.s^{-1}

Table 1: Properties of the water-ethanol mixture at 60.88% water by mass and 22.5°C.

 For a temperature difference of 1°K we obtain e_{opt} =0.4 mm, and the corresponding separation for a length of 0.53 m is: 0.37, which is a very large separation.

Figure 2 shows the variation in separation as a function of cell thickness, demonstrating the existence of an optimum thickness value.

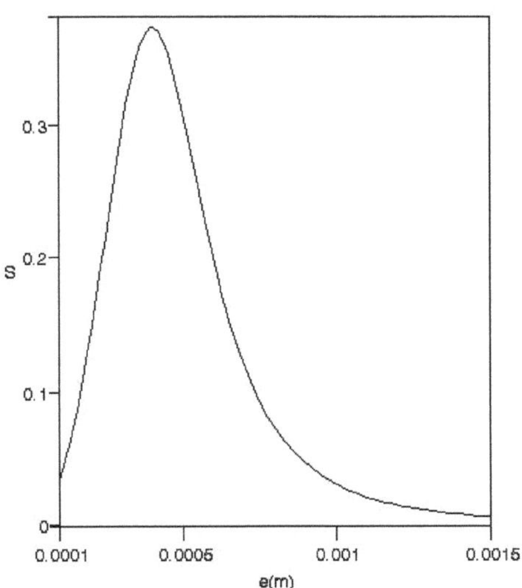

Figure 2: Separation as a function of cell thickness for L=0.53m and
$$\Delta T = 1°K$$

Given this large separation, the hypothesis of replacing $C(1-C)$ with $C_0(1-C_0)$ is questionable.

4. **Results for large separations :**

Majumdar [21] solved the problem with variable C and gave the general form of the separation, also specifying the associated separation time.

He obtained the following relationship giving the separation as a function of a parameter q :

$$\Delta C = \frac{\left(q^{C_0}-1\right)\left(q^{1-C_0}-1\right)}{(q-1)} \tag{30}$$

      With :

21

$$q = e^{(H_1/K)H}$$

$$\frac{H_1}{K} = \frac{\dfrac{\beta_T\,\rho g D_T\,\Delta T^2 e^3}{6!\nu D}}{\dfrac{\beta_T^{\,2}\rho g^2\,\Delta T^2 e^7}{9!\nu^2 D} + e\rho D} \tag{31}$$

The separation time (relaxation time) is given by :

$$\tau = \frac{9!\,H^2\nu^2 D}{(\pi\,\Delta T g\,\beta_T^{\,3})^2} \tag{32}$$

For a temperature difference of 1°K and a thickness of 0.4 mm, we obtain a relaxation time of 36601 hours, i.e. about 4 years!

Further separation can be achieved by increasing the temperature difference, but this will reduce the optimum thickness, which is more difficult to achieve, and the volume of fluid to be separated becomes very small.

Table 2 shows the optimum thickness for different ΔT values, with the corresponding separation for a 0.53 m length, and the separation time.

ΔT	1°K	5°K	10°K	15°K
Optimum thickness	0.40mm	0.23mm	0.18mm	0.16mm
Separation	0.3731243264	0.9958086195	0.9999990887	1
Relaxation time	36601.2 hours	36601.2 hours	36601.2 hours	36601.2 hours

Table 2: Optimum separation in a vertical cell 0.53 m long.

This is due to the fact that, by replacing the optimum thickness by its value in the relaxation time expression (32), the temperature difference is simplified and the relaxation time becomes independent of ΔT , which can be explained by the fact that increasing the temperature difference increases convection, but decreasing the thickness decreases it, so basically we have the same convection intensity and therefore the same relaxation time. If we replace the thickness e by the optimum thickness e_{opt} in equation (32), we obtain the relaxation time for the optimum thickness:

$$\tau = \frac{2H^2}{D\pi^2} \tag{33}$$

The vertical cell can therefore achieve complete separation, but the thicknesses required are very small and difficult to achieve, and the volume of the mixture is very small. In addition, the relaxation time is very long due to the low thickness.

In practice, it is impractical to achieve a thickness of less than 1.5 mm along a cell length of more than 50 cm.

Table 3 shows the variation in separation and relaxation time as a function of temperature difference for a thickness of 1.5 mm and a length of 530 mm.

ΔT	1°K	5°K	10°K	15°K
Separation	0.006059	0.006063	0.006064	0.006064
Relaxation time	13.5 hours	5.3 hours	1.3 hours	0.6 hours.

Table 3: Variation of separation and relaxation time as a function of temperature difference for water-ethanol mixture with 60.88% water by mass for H=53cm.

It can be seen that the relaxation time decreases rapidly when ΔT is increased, which can be explained by the fact that when ΔT is increased, the convection velocity increases.

Separation remains practically constant as ΔT varies. This is because the second term in the denominator of $\frac{H1}{K}$ in equation (31) is negligible compared with the first term for the 1.5 mm thickness:

$$\frac{\beta_T^2 \rho g^2 \Delta T^2 e^7}{9! v^2 D} = 8.21 \times 10^{-7} \times \Delta T^2 \quad et \quad e\rho D = 0.606 \times 10^{-9}$$

Therefore, in the expression of $\dfrac{H_1}{K}$, ΔT simplifies and the separation becomes independent of . ΔT

5. Conclusion:

In this chapter, we discuss in detail the study of **thermogravitation** in a vertical cell, a classical model used for the separation of binary mixtures under the effect of a thermal gradient. Building on the theoretical work of **Onsager, Furry and Jones**, we have developed a rigorous analytical approach to solving the equations governing natural convection and thermodiffusion phenomena in a cavity of simple geometry, but of great practical interest.

The results obtained show that, under the effect of a temperature gradient applied to the vertical walls, it is possible to generate a separation of the constituents of a binary mixture. The temperature field, combined with the concentration field, can be used to determine the optimum conditions for maximizing separation. We have been able to demonstrate that there is an optimum cavity thickness for which species separation is maximized. This thickness is directly related to the thermophysical properties of the mixture under study, notably the **thermal Rayleigh number** and the **Soret coefficient**, which describes the intensity of the thermodiffusion effect.

However, although this vertical configuration enables significant separations to be achieved, our analyses reveal certain limitations inherent in the system. Indeed, the optimum cavity thickness is often very small (of the order of a few tenths of a millimetre), which complicates its practical realization in experimental or industrial systems. What's more, when the thickness exceeds this optimum value, separation drops considerably, and the time required to achieve stable separation, also known as the **relaxation time**, becomes very long, reaching several thousand hours in some cases. This limits the efficiency and speed of separation processes in this type of configuration.

Another important aspect highlighted in this study is the **non-linearity** of convection and thermodiffusion effects when moving away from optimum conditions. At high Rayleigh numbers, convection becomes too intense, rapidly homogenizing the mixture and cancelling out the effects of thermodiffusion. This demonstrates the crucial importance of optimum coupling between the two phenomena to achieve efficient separation.

Finally, although the theory and analytical solutions developed in this chapter provide robust and useful results for simple configurations, they cannot be applied directly to more complex or realistic systems. In particular, effects related to variations in geometric structure, non-linearity of thermophysical properties, and the influence of edge effects require more sophisticated approaches, such as numerical simulations.

Thus, in the following chapters, we will focus on alternative configurations, including **horizontal cavities**, which could offer significant practical advantages. We will also explore numerical approaches to modeling flows and mass and heat transfers in porous systems saturated with binary mixtures, with a view to optimizing separation processes under more varied experimental conditions.

Chapter 3

Analysis of Thermo-Gravitation Phenomena in Horizontal Layers: Towards Efficient Separation of Fluid Mixtures

1. Introduction :

Understanding the separation mechanisms of fluid mixtures is essential in many fields of engineering and applied science, including water treatment, hydrocarbon separation and chemical process management. Among the various separation methods, thermogravitation stands out for its ability to exploit both thermal and gravitational gradients to promote separation of the constituents of a mixture. This phenomenon, which combines thermodiffusion (or the Soret effect) and natural convection, optimizes the separation of species in multi-component systems.

In this chapter, we focus on the study of thermogravitation in a flat layer subjected to horizontal heat flow. This configuration offers unique features that differ from conventional systems such as vertical cells. Indeed, the horizontal arrangement of the heat flow enables better management of temperature gradients and offers practical advantages in terms of separation optimization. The geometric simplicity of the flat layer, combined with the effect of gravity, also facilitates modeling and theoretical analysis of the phenomenon.

The main objective of this chapter is to develop an analytical approach to modeling thermogravitation phenomena in this type of configuration. We begin by establishing the governing equations that describe steady-state fluid behavior, integrating both convection and thermodiffusion effects. Considering the laws of conservation of mass, momentum and energy, we will formulate the equations needed to describe heat and mass transfer within the layer.

We will also examine the influence of boundary conditions and dimensional parameters, such as the Rayleigh number, on the separation process. The Rayleigh number, which quantifies the ratio between buoyancy and viscous forces, plays a crucial role in flow stability and species separation. By analyzing different flow configurations and assessing the effects of temperature variations, we will seek to identify the optimum conditions for maximizing constituent separation.

Numerical simulations will also be carried out to validate our analytical results and explore operating regimes that cannot be easily captured by analytical models. The use of advanced computational methods, such as the finite element method, will enable us to examine the complex interactions between the various physical phenomena at work in the system.

Finally, we will discuss the practical implications of these results for the development of more efficient separation technologies. By identifying the challenges associated with implementing these concepts in industrial applications, we will highlight future directions for thermogravitation research, particularly with regard to optimizing existing systems and exploring new geometric configurations.

In sum, this chapter aims to provide an in-depth understanding of thermogravitation separation mechanisms in a flat-layer configuration, while laying the foundations for future industrial applications.

2. Problem presentation :

We consider a rectangular cavity of length L' and height H' such that the aspect ratio $A = L'/H' \gg 1$ (Figure 1). The vertical concentration gradient induced by the Soret effect results from the imposition of a constant, uniform heat flux q' on the horizontal walls. The vertical walls of the cavity are assumed to be adiabatic and impermeable. The porous medium is assumed to be homogeneous, isotropic, obeying Darcy's model and saturated by a binary, incompressible fluid.

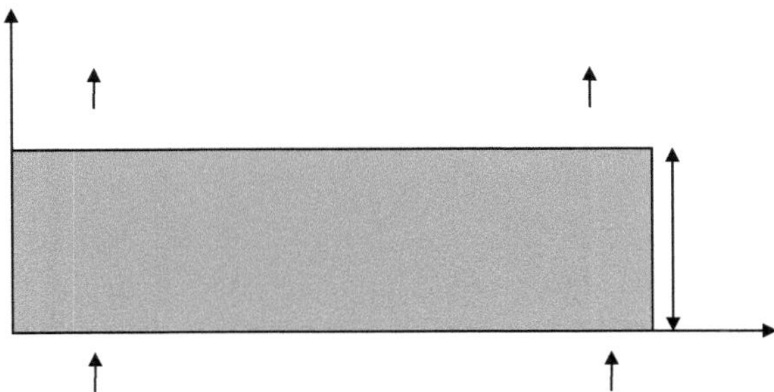

Figure 1: Rectangular cell of length L' and height H' subjected to heat flow :

Note: Quantities marked with " ' " are dimensional quantities.

First, we'll look at the dimensional equations governing the behavior of the mixture. We will then put these equations into adimensional form to simplify the number of problem parameters. The adimensional system will then be solved using a numerical computation program "Comsol" which employs the finite element method, and the results obtained will be compared with the analytical results, the results obtained in a vertical cell and finally with those obtained by Vasseur et al [33].

3. **Mathematical formulation :**

The cavity used is described above (see Fig. 1). The aim of this investigation is to determine the velocity, temperature and concentration fields as a function of the problem parameters.

We present Darcy's equations governing the behavior of a fluid in a porous medium, and the simplifying assumptions that allow us to characterize its motion within the cavity.

Therefore, to solve this problem, we take into account the following assumptions:

- The fluid studied is considered to be Newtonian and incompressible.
- The work of pressure forces and the work due to viscosity forces will be neglected.
- The flow is assumed to be two-dimensional.
- We are in steady state.
- We place ourselves in the Boussinesq approximation

$$\rho = \rho_0 \left[1 - \beta_T \left(T - T_0\right) - \beta_C \left(C - C_0\right)\right] \tag{1}$$

Where T and C respectively represent the temperature and mass fraction of the fluid at a point in the system, and , ρ_0 T_0 and C_0 respectively symbolize the density, temperature and concentration in the initial state. β_T and β_C are respectively the thermal and mass expansion coefficients.

$$\left(\beta_T = -\frac{1}{\rho_0}\frac{\partial \rho}{\partial T}\bigg|_{P,N} > 0 \quad et \quad \beta_C = -\frac{1}{\rho_0}\frac{\partial \rho}{\partial C}\bigg|_{P,T} < 0\right) \tag{2}$$

1. **Dimensional governing equations :**

For the porous medium, the laminar flow of homogeneous fluids is governed by Darcy's law, so that the effect of inertial forces can be neglected. So the system of dimensional equations of the problem is :

$$\begin{cases} \nabla^2 \Psi' = -\dfrac{gK\beta_T}{v}\dfrac{\partial}{\partial x'}\left(T' + \dfrac{\beta_C}{\beta_T}C\right) & (4) \\[2ex] (\rho c)_p \dfrac{\partial T'}{\partial t'} + (\rho c)_f \overrightarrow{V'}.\overrightarrow{\nabla}T = k\nabla^2 T' & (5) \\[2ex] \phi\dfrac{\partial C}{\partial t'} + \overrightarrow{V'}\overrightarrow{\nabla}C = D\nabla^2 C + D_T C(1-C)\nabla^2 T' & (6) \end{cases}$$

Where $\overrightarrow{V'}$ is the Darcy velocity, $\overrightarrow{g}$ is the gravitational constant, v is the kinematic viscosity, $(\rho c)_f$ et $(\rho c)_p$ are the heat capacities of the fluid and porous medium respectively, k is the thermal conductivity, D et D_T are the mass diffusion coefficient and thermodiffusion coefficient respectively, K is the permeability, ϕ is the porosity of the porous medium and Ψ' is the current function.

In heat transfer problems it is common to eliminate pressure by using the current function formulation Ψ'. In the following form :

$$u' = \frac{\partial \Psi'}{\partial y'} \quad \text{and} \quad v' = -\frac{\partial \Psi'}{\partial x'} \tag{7}$$

The corresponding boundary conditions :

$$x' = \pm L', \qquad \Psi' = 0, \qquad \frac{\partial T'}{\partial x'} = \frac{\partial C}{\partial x'} = 0 \tag{8}$$

$$y = \pm H', \qquad \Psi' = 0, \qquad \frac{\partial T'}{\partial y'} = -\frac{q'}{k} \tag{9}$$

$$\frac{\partial C}{\partial y'} = -\frac{D_T}{D}C(1-C)\frac{\partial T'}{\partial y'} \tag{10}$$

q' is the heat flux applied to horizontal walls.

2. Scaling parameters :

The above equations have been established in dimensional form. To simplify calculations, we convert them to dimensionless form.

In order to make the previous equations dimensionless, the H' dimension of the cavity was chosen as the reference length. The ratio $\dfrac{a}{H}$ was used to dimension the velocity. Temperature and concentration are adimensionalized respectively with respect to the temperature $\Delta T'$ and concentration ΔC gaps between the two horizontal walls of the cavity.

The scaling parameters chosen are as follows:

$$(x,y) = \frac{(x',y')}{H'}, \quad (u,v) = \frac{(u',v')\,H'}{a}\ , \ S = \frac{C}{\Delta C}$$

$$t = t'\frac{a}{H'^2}, \qquad \Psi = \frac{\Psi'}{a}, \qquad \varepsilon = \frac{\phi}{\sigma}$$

$$T = \frac{(T'-T'_0)}{\Delta T'}, \qquad\qquad \Delta T' = \frac{q'H'}{k}$$

With a the thermal diffusivity of the fluid: $a = k/\rho C$

$\Delta C = -C_0 \Delta T'(1-C_0)\ D_T/D$ and $\sigma = \dfrac{(\rho c)_p}{(\rho c)_f}$ is the ratio of heat capacities.

3. Dimensionless governing equations and boundary conditions :

If we replace the dimensionless quantities in equations {(4), (5) and (6)} we obtain the following dimensionless system:

$$\nabla^2\Psi = -Ra\left(\frac{\partial T}{\partial x} + \varphi\frac{\partial S}{\partial x}\right) \tag{11}$$

$$\frac{\partial T}{\partial t} + u\frac{\partial T}{\partial x} + v\frac{\partial T}{\partial y} = \nabla^2 T \tag{12}$$

$$\varepsilon\frac{\partial S}{\partial t} + u\frac{\partial S}{\partial x} + v\frac{\partial S}{\partial y} = \frac{1}{Le}\left(\nabla^2 S - \nabla^2 T\right) \tag{13}$$

This problem depends on 5 dimensionless parameters: the thermal Rayleigh number Ra , the solutal Rayleigh number Ra_S , the Lewis number Le , the normalized porosity ε and the aspect ratio A . These parameters are given by :

$$Ra = \frac{g\beta_T K\Delta T' H'}{a\nu}, \qquad Ra_S = \frac{g\beta_N K\Delta C H'}{D\nu}$$

$$Le = \frac{a}{D}, \qquad A = \frac{L'}{H'}, \qquad \varepsilon = \frac{\phi}{\sigma}$$

The thermal and solutal Rayleigh numbers are related by the following expression : $Ra_S = Ra\varphi Le$ where φ is the buoyancy rate defined by : $\varphi = \dfrac{\beta_N \Delta C}{\beta_T \Delta T'}$

The boundary conditions corresponding to the system of equations {(11), (12), (13)} are :

$$\text{In } y = 0 : \quad \Psi = 0 \qquad\qquad \frac{\partial T}{\partial y} = \frac{\partial S}{\partial y} = -1 \qquad\qquad (14)$$

$$\text{In } y = 1 : \quad \Psi = 0 \qquad\qquad \frac{\partial T}{\partial y} = \frac{\partial S}{\partial y} = -1 \qquad\qquad (15)$$

$$\text{At } x = 0 : \quad \Psi = 0 \qquad\qquad \frac{\partial T}{\partial y} = \frac{\partial S}{\partial y} = 0 \qquad\qquad (16)$$

$$\text{In } x = A : \quad \Psi = 0 \qquad\qquad \frac{\partial T}{\partial y} = \frac{\partial S}{\partial y} = 0 \qquad\qquad (17)$$

4. Analytical resolution :

In the previous paragraph, we established the basic equations governing thermosolutal convection. These equations form a system of coupled, non-linear differential equations that are difficult to solve analytically. However, in the limiting case of large porous layers $(A \gg 1)$, approximate analytical solutions are possible.

The aim of this section is to present the approximate analytical method used to solve the problem posed, and to present the assumptions that allow the problem to be solved more easily, leading to a simplified solution.

The governing equations can be simplified using the parallel flow hypothesis (used in the past by several authors, including Sen et al. [16], Elhajjar, Mojtabi and

Marcoux [12]), who consider that the flow generated in the core of an infinite cavity $(A \gg 1)$ becomes parallel to the walls in the central region. The flow reverses by 180° at the ends of the cavity. Temperature and concentration profiles then depend on the transverse coordinate.

- The present problem can be considered for an elongated cavity ($A \to \infty$). In this case, as discussed by Elhajjar, Mojtabi and Marcoux. [12] and other authors, the flow can be considered as parallel in the central region of the cavity. The current function then depends only on y :

$$\Psi(x, y) \approx \Psi(y) \qquad (18)$$

- In the central portion of the cavity, the fluid moves at constant speed and is subject to the uniform heat flow $q\,'$ along its entire path. The temperature distribution function therefore varies linearly along x. Since the variation along y is unknown, we introduce the function $f(y)$ such that :

$$T(x, y) = b\, x + f(y) \qquad (19)$$

- Similarly, the concentration distribution function can be written as :

$$S(x, y) = m\, x + g(y) \qquad (20)$$

Where b and m are respectively the temperature and concentration gradients along the x axis. These gradients are introduced to take into account the influence of the cavity edges on the rest of the flow.

4.1 Application to governing equations :

As we saw earlier, both temperature and concentration are functions of x and y. The current function depends only on y. Moreover, we are in a steady state. The dimensionless equations (11) (12) and (13) can be written as follows:

$$\frac{d^2\Psi}{dy^2} = -\,Ra(b + \varphi\, m\,) \qquad (21)$$

$$b\frac{d\Psi}{dy} = \frac{\partial^2 f}{\partial y^2} \qquad (22)$$

$$m\frac{d\Psi}{dy} = \frac{1}{Le}(g'' - f'') \qquad (23)$$

To solve this system of equations with the associated boundary and physical conditions, we used the MAPLE symbolic calculation software (see Appendix 1).

4.2 Problem-solving approach :

4.2.1   Determining the current function :

Integrating equation (21) we find :

$$\Psi = -\frac{1}{2}Ra(b + \varphi m)y^2 + C_1 y + C_2$$

C_1 *et* C_2 are integration constants to be determined by the boundary conditions.

Visit $y = 0$, $\Psi(y) = 0$

Visit $y = 1$, $\Psi(y) = 0$

Finally, the current function equation is written :

$$\Psi = \Psi_0(y - y^2) \tag{24}$$

With
$$\Psi_0 = \frac{1}{2}Ra\ (b + \varphi m) \tag{25}$$

4.2.2   Determination of temperature and concentration distribution functions :

By integrating equation (22) twice:

$$f(y) = \frac{1}{2}\Psi_0 b y^2 - \frac{1}{3}\Psi_0 b y^3 + C_3 y + C_4$$

C_3 and C_4 are two constants to be determined by the boundary conditions (see Appendix 1):

$$f(y) = \frac{1}{2}\Psi_0 b y^2 - \frac{1}{3}\Psi_0 b y^3 - y \tag{26}$$

In the same way for the concentration, we integrate equation (23) and replace $f(y)$ by its expression found above:

$$g(y) = \left(\Psi_0 mLe + \Psi_0 b\right)\left(\frac{1}{2}y^2 - \frac{1}{3}y^3 - \frac{1}{12}\right) - y + \frac{1}{2} - \frac{1}{2}mA \tag{27}$$

4.2.3   Determination of temperature gradient b and mass fraction m :

Under steady-state conditions, the horizontal concentration profile stabilizes, so there is no mass flow through any vertical section of the cell:

$$\int_0^1 (u(y)g(y)Le - m + b)dy = 0 \tag{28}$$

The solution to this equation gives the expression for the concentration gradient m :

$$m = -\frac{Le\Psi_0^2 b - 5Le\Psi_0 - 30b}{Le^2\Psi_0^2 + 30} \tag{29}$$

Using the same procedure to find the expression for the temperature gradient b, there is no longer any heat flow through any vertical section of the cell:

$$\int_0^1 (u(y)f(y) - b)dy = 0$$

So..: $$b = \frac{5\Psi_0}{\Psi_0^2 + 30} \tag{30}$$

By replacing the values of m and b by their expressions in relation (25), we obtain a polynomial equation of degree 5 in . Ψ_0

$$2\Psi_0^5 Le^2 + (60Le^2 + 60 - 5RaLe^2)\Psi_0^3 + (-150Ra - 150Ra\phi Le + 1800 - 150Ra\phi)\Psi_0 = 0$$

This equation has the following five solutions: $\tag{31}$

$$\Psi_0 = 0$$

$$\Psi_0 = \pm\sqrt{5}\left(\frac{RaLe^2 - 12 - 12Le^2 \pm}{\sqrt{Ra^2Le^4 + 24RaLe^2 - 24RaLe^4 + 144 - 288Le^2 + 144Le^4 + 48Le^3 Ra\varphi + 48Le^2 Ra\varphi}}\right)^{1/2}$$

So the approximate analytical solution of the problem is :

$$\begin{cases}
\Psi = \Psi_0(y - y^2) \\[2mm]
T_0 = bx + \Psi_0 by^2\left(\dfrac{1}{2} - \dfrac{y}{3}\right) - y \\[2mm]
c_0 = mx + \dfrac{1}{2}\Psi_0 mLey^2 - \dfrac{1}{3}\Psi_0 mLey^3 + \dfrac{1}{2}\Psi_0 by^2 - y - \dfrac{1}{12}\Psi_0 b - \dfrac{1}{12}\Psi_0 mLe + \dfrac{1}{2} - \dfrac{1}{2}mA \\[2mm]
\Psi_0 = \pm\sqrt{5}\left(\dfrac{Ra_T Le^2 - 12 - 12Le^2 \pm}{\sqrt{Ra_T^2 Le^4 + 24Ra_T Le^2 - 24Ra_T Le^4 + 144 - 288Le^2 + 144Le^4 + 48Le^3 Ra_T\varphi + 48Le^2 Ra_T\varphi}}\right)^{1/2}
\end{cases} \qquad (32)$$

5. Separation:

Separation is defined as the difference in the mass fractions of the heaviest component between the left and right ends of the cell $S = mA$, so it is easy to find that maximum separation is obtained for an optimum value of Ra :

$$Ra = \frac{1}{Le^2(10Le^2 + 197Le - 253)}(40Le^4 - 2400Le^2 + 2000Le^3 - 400Le + 440$$
$$+ 40\sqrt{121 - 220Le + 1816Le^2}\,)^{1/2}$$

This value will be noted . Ra_{opth}

Figure 2 shows the variation of the separation as a function of the Rayleigh number . Ra

This maximum corresponds to the optimum coupling between convection and thermodiffusion. If Ra is small, convection is weak and separation is mainly due to thermodiffusion. If Ra is large, convection is intense and the flow mixes the constituents of the mixture, resulting in low separation. It should be noted that, depending on the reference value chosen to render the concentration dimensionless, the separation may be greater than 1.

For , $Le = (2,50,100)$ $\varphi = 0.4$, Figure 2 shows the corresponding optimum Rayleigh numbers ($Ra_{opt} = 7.85$ for , $Le = 2$ $Ra_{opt} = 1.07$ for $Le = 50$ and $Ra_{opt} = 0.56$ for $Le = 100$), and the corresponding maximum separations (S= 5.76 for Le= 2 , S= 4.65 for Le= 50 and 4.61 for Le= 100).

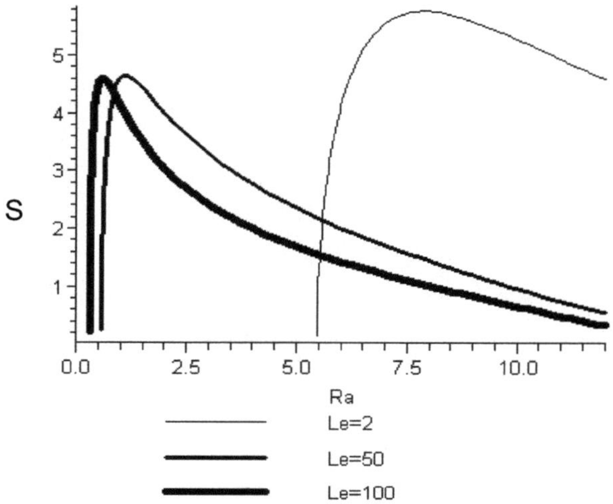

Figure 2. separation curve for , , , . $Le = 2\ 50\ 100$ $\varphi = 0.4$

5.1 Separation obtained in a vertical cell compared to that obtained in the Rayleigh-Bénard configuration:

To demonstrate the practical benefits of the horizontal configuration, we compare it with the vertical configuration. A thermogravitation study is carried out in a vertical cell of height H and thickness e . The vertical walls of the cavity are maintained at different, constant temperatures. We use the same assumptions as for a horizontal cell.

By solving the equations of the problem with the corresponding boundary conditions and the assumptions adopted. In this case, the temperature gradient is $m = (10LeRa)/(Le^2Ra^2 + 120)$. Looking at this expression, we see that there is a maximum separation for $Ra = \sqrt{120}/Le$. This value will be noted Ra_{optv} . [12].

Replacing the optimum values of Ra in the expressions for vertical and horizontal separations, we obtain almost the same maximum separation values. For example, if we take Le=100, we find $S_{vmax} = 4.56$ et $S_{hmax} = 4.60$ and therefore $S_{vmax} \approx S_{hmax}$. Thus, the maximum separations in the vertical and horizontal cases are equal. To demonstrate the advantage of the horizontal configuration, we'll calculate the ratio of optimal Rayleigh numbers in both cases:

$r = \text{Ra}_{opth}/\text{Ra}_{optv}$

$$= \frac{0.09128709291\left(40Le^4 - 2400Le^2 + 2000Le^3 - 400Le + 440 + 40\sqrt{121 - 220Le + 1816Le^2}\right)}{Le(10Le^2 + 197Le - 253)}$$

For example, if we consider the Lewis values (Le=2, 50 and 100), we can easily find the numerical values of this ratio.

For Le=2 we get $r = 2.525138846$, for Le=50 we get $r = 26.07267065$ and for Le=100 we get $r = 45.67244110$. So every time we increase Le, the ratio also increases, so we've checked that $r > 1$ so $\text{Ra}_{opth} > \text{Ra}_{optv}$. So the optimum height of the horizontal cell is greater than the width of the vertical column, which is easier to achieve in experiments, and allows us to obtain a greater quantity of separated fluid. Another disadvantage of the vertical cell is the rapid drop in separation as the Rayleigh number increases from the optimum value. On the other hand, in the case of the horizontal cell, we see a slow decrease in separation as the Rayleigh number increases (Fig. 3), so a Rayleigh number higher than the optimum can be used without any significant loss in the degree of separation.

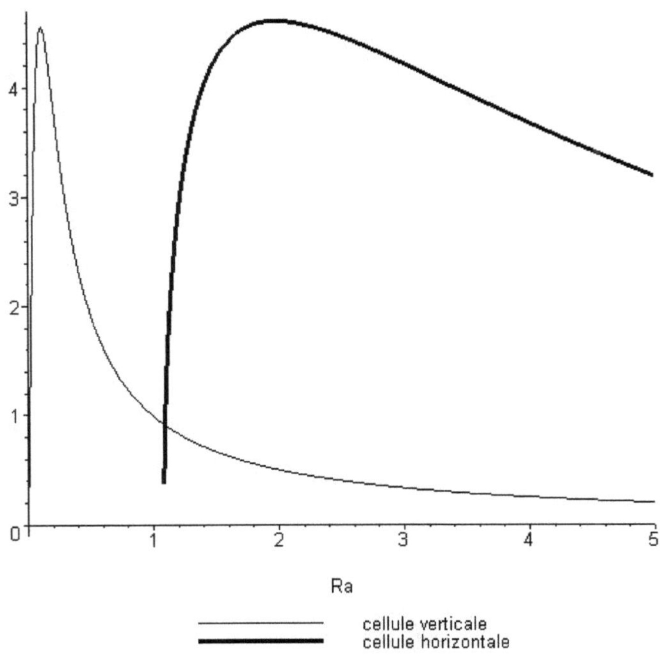

Figure 3. Separation in horizontal and vertical cells for , $Le = 100$ $\varphi = 0.4$ and A=10

6. Digital simulation :

Equations (3.36),(3.37) and (3.38) ,with boundary conditions were solved numerically using a finite element method (industrial code Comsol) with a quadrangle mesh better suited to the rectangular cell shape used, and spatial resolution 150 x 30. We found good agreement between analytical and numerical results as regards separation for $Le = 100$, , . $\varphi = 0.1$ $A = 10$

6.1 Introducing COMSOL :

COMSOL is a powerful, interactive environment for modeling all kinds of scientific problems based on partial differential equations (PDEs). With COMSOL you can easily extend conventional models for one type of physics into multiphysics models that solve coupled physics phenomena, and do so simultaneously. By employing integrated modes of physics it is possible to establish models by defining the appropriate physical quantities such as material properties, charges, stresses, sources and flows, rather than by defining fundamental equations.

COMSOL then internally compiles a set of PDEs representing the entire model. The power of COMSOL can be accessed as a stand-alone product via a flexible graphical user interface, or by hand, by programming in the MATLAB language.

As already mentioned, the fundamental mathematical structure in COMSOL is a system of partial differential equations. There are three ways of describing PDEs by the following mathematical modes:

- "Coefficient form", suitable for linear or near-linear models.
- "General form", suitable for non-linear models.
- "Weak form", for models with PDEs at boundaries, edges, or points, suitable for models using terms with space and time derivatives.

Using these application modes, various types of analysis can be performed including:

- Eigenvalue analysis.
- Stationary and time-dependent analysis.
- Linear and non-linear analysis.

When solving PDEs, COMSOL employs the finite element method. The software performs finite element analysis with mesh adaptation and error control using a variety of numerical solvers. PDEs provide the basis for modeling a wide variety of scientific phenomena. As a result, COMSOL can be used in a wide range of applications, such as :

Heat transfer, fluid dynamics, acoustics, biosciences, diffusion, optics, flow in porous media, quantum mechanics, transport phenomena...

To numerically simulate the problem of thermogravitational diffusion in the horizontal cell, we considered the system of basic equations already solved analytically.

6.2 Comparison of analytical and numerical results :

Table 1 and Figure 4 show the separation results obtained analytically and numerically.

Ra	separation (digital)	separation (analytical)	Analytical numerical deviation (%)
1.5	4.2486725	4.252121	-0.081100702
2	4.6257725	4.609307	0.357222897
3	4.223495	4.212462	0.261913342
4	3.6816775	3.673587	0.220234338
6	2.7610025	2.753509	0.272143654
8	2.0116375	2.001598	0.501574242
10	1.31356	1.296308	1.330856556
12	0.5269825	0.500102	5.375003499
14	0.15694	0.188546	-16.76301804
18	0.03599	0.086475	-58.38103498
26	0.00707	0.039328	-82.02298617
30	0.0130875	0.030073	-56.48089649
40	0.018465	0.018093	2.056043774
50	0.0194075	0.012428	56.1594786
60	0.0191375	0.009216	107.6551649

| 70 | 0.018505 | 0.007187 | 157.4787811 |

Table 1: Variations of numerical separation results with Rayleigh number for .
$(Le = 2 \; et \; \varphi = 0.4)$

From the results in the table above, we can see that the difference between analytical and numerical results is very small for low Rayleigh numbers.

Figure 4 shows the analytical and numerical separation results we have found and those obtained by Vasseur et al [33].

We find that the numerical separation results are in good agreement with our analytical results for low Rayleigh numbers. The agreement is less good for high Rayleigh numbers. This is due to the fact that higher Rayleigh numbers increase convection velocities, so edge-turning effects are more important and the assumption of linearity of the concentration field is no longer entirely valid. The results of Vasseur et al [33] show in Fig. 4 shows that there is a considerable discrepancy between the numerical and analytical results.

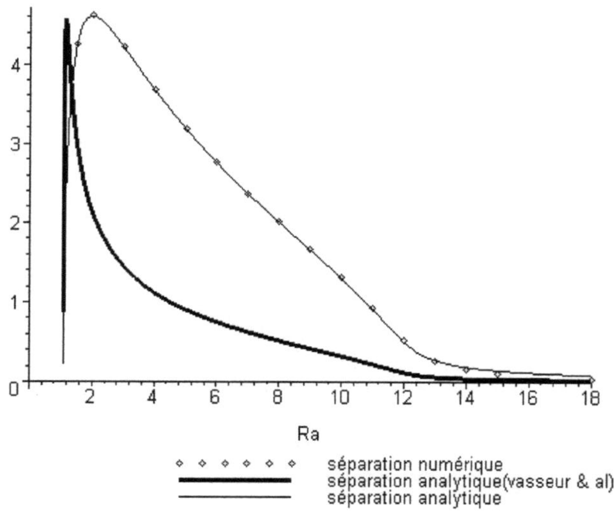

Figure 4: Analytical and numerical separation curves for , $Le = 100 \; \varphi = 0.1$

By observing the concentration fields obtained numerically, we have noticed that the edge effects neglected during the analytical resolution of the problem, and the assumption of linearity of the concentration field along x, are no longer entirely acceptable for high Rayleigh numbers. Figure 6 shows the variation of the concentration field as a function of x for the cell studied in this chapter, demonstrating that the concentration field along x is perfectly linear. We can see that edge effects are very important and that the concentration field is no longer linear, but the slope of the concentration profile changes abruptly near the edges.

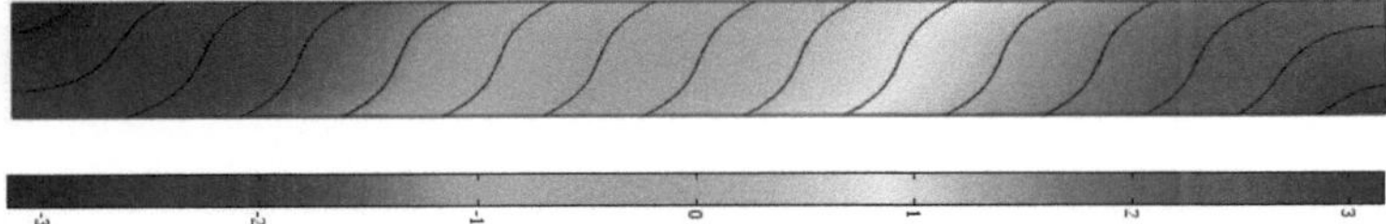

Figure 5: Isoconcentrations in the cell for Ra=8, Le=2

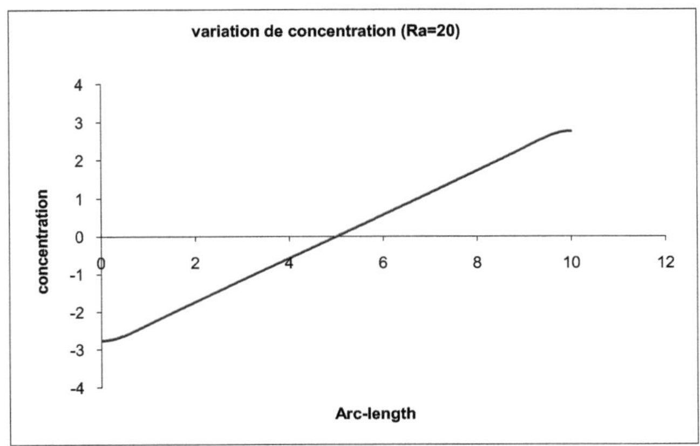

Figure 6: Variation of the concentration field as a function of x for a horizontal cell.

Figures 7 and 8 show, respectively, the isotherms in the binary mixture and the variation in temperature along a line linking the two extreme points placed at mid-height of the cell.

41

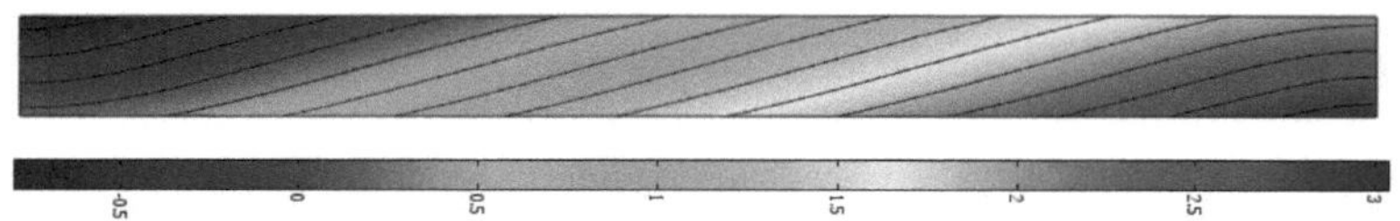

Figure 7: Isotherms in the cell for Ra=8, Le=2

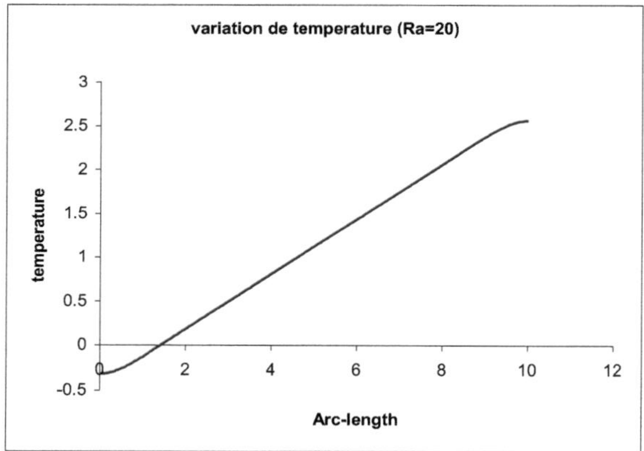

Figure 8: Variation of the temperature field as a function of x for a horizontal cell

Figure 9 shows isoconcentrations and streamlines for Ra=20. A single-cell flow is observed in this case due to thermodiffusion, with the denser component of the mixture migrating towards the colder wall, i.e. the upper wall, and causing a destabilizing effect compared to the pure thermal case. While the other component moves towards the warmer, lower wall ($\varphi > 0$), the single-cell flow advects one of the mixture's components towards the right-hand side of the cavity and the other towards the left-hand side. This results in a high concentration of one species in the left-hand side of the cell and a high concentration of the other species in the right-hand side, leading to horizontal stratification of the concentration field and separation of the species in the mixture.

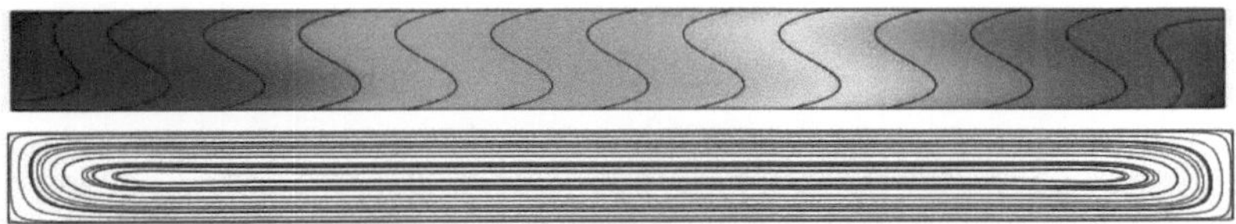

Figure 9: Isoconcentrations and current lines for : Ra=20, $\varphi = 0.4$ and Le=2 (t=10000).

Figure 10 shows the variation of the concentration field as a function of the Rayleigh number Ra, for a cell of dimensions 2mm x 20mm. The lines represent isoconcentrations and the colors the intensity of the water mass fraction.

At low Rayleigh numbers, e.g. Ra=4, the isoconcentrations are horizontal, and at Ra=6 they are almost horizontal, showing that separation is mainly due to thermodiffusion. As the Rayleigh number increases, the convective velocity increases and we get closer and closer to a horizontal separation. As Ra increases further, convection becomes more important than thermodiffusion, the curvature of isoconcentrations increases and separation decreases.

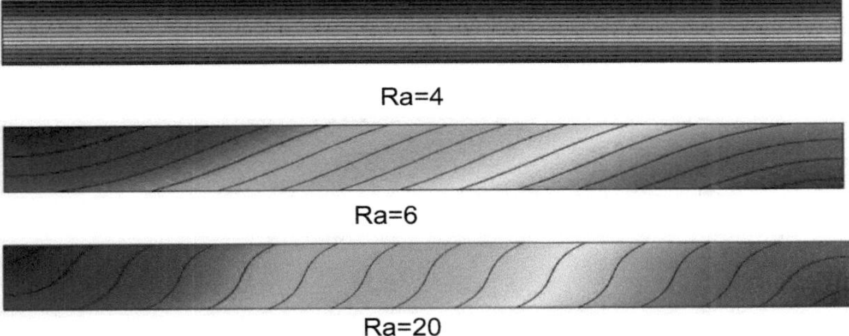

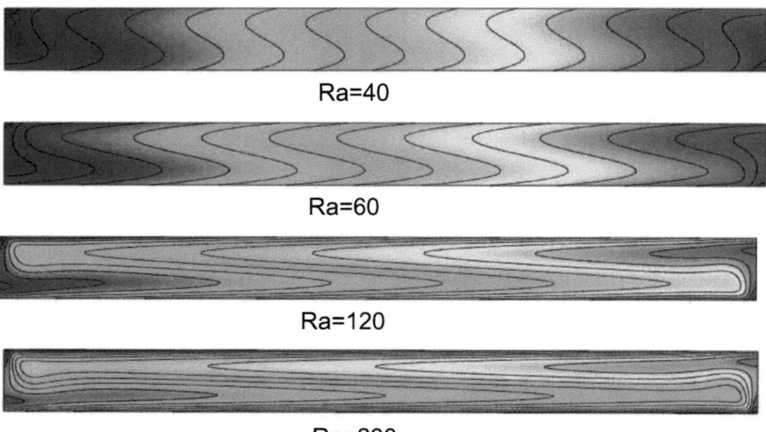

Figure 10: Variation of isoconcentrations as a function of Rayleigh number Ra for

(Le= 2 and)$\varphi = 0.4$

7. Linear stability analysis of single-cell flow :

To study the stability of the single-cell solution, we write the equations introducing perturbations of the current function Ψ , temperature T and mass fraction S.

$$\Psi = \Psi_0 + \Phi \ , \ T = T_0 + \theta \ , \ S = S_0 + s$$

Neglecting the second-order terms, we obtain the linear equations whose unknowns are the disturbances, after which we develop disturbances in normal modes:

$$(\Phi,\theta,s) = (\Phi(y),\theta(y),s(y))e^{(Ikx+\sigma t)} \ ,$$

We obtain the following system of equations:

$$
\left\{
\begin{array}{l}
(D^2 - k^2)\Phi(y) = -IkRa(\theta(y)+\varphi s(y)) \\[3mm]
\sigma\theta(y) + Ik\Psi_0(1-2y)\theta(y) + b\Phi'(y) - Ik(\Psi_0 by - \Psi_0 by^2 - 1)\Phi = (D^2 - k^2)\theta \\[3mm]
\varepsilon\sigma s(y) + Ik\Psi_0(1-2y)s(y) + m\Phi'(y) - Ik(\Psi_0 mLey - \Psi_0 mLey^2 + \Psi_0 by - \\[2mm]
\Psi_0 by^2 - 1)\Phi = \dfrac{1}{Le}(D^2 - k^2)(s(y) - \theta(y))
\end{array}
\right.
$$

Where , $D = d/dy$ k is the wave number in the horizontal direction, $\sigma = \sigma_r + I\sigma_i$ is the time amplification rate of the disturbance, and $. I^2 = -1$

The corresponding boundary conditions are

$$\Phi = 0, \quad \frac{d\theta}{dy} = 0, \quad \frac{ds}{dy} = 0 \quad en \quad y = 0, 1$$

The resulting linear problem is solved by Galerkin's method to order 5 by expanding the perturbations as follows (see Appendix 2):

$$\Phi = \sum_{i=1}^{N} a_i \sin(i\pi y) \quad \theta = \sum_{i=0}^{N-1} b_i \cos(i\pi y) \quad s = \sum_{i=0}^{N-1} c_i \cos(i\pi y)$$

Critical Rayleigh number values were obtained for stationary and unsteady bifurcations. The critical Rayleigh number leading to a stationary bifurcation is always greater than the critical Rayleigh number leading to the Hopf bifurcation. For this reason, in this study we mainly present the values of the wave number k_C , the critical Rayleigh number Ra_C and the critical pulsation ω_C associated with the Hopf bifurcation.

The linear stability results give $Ra_c = 265.51$, $k_c = 4.68$ et $\omega_c = 97.53$ for $Le = 2$ et $\varphi = 0.4$ (see Appendix 2) in an infinite cell. However, in the numerical study, we were unable to find the same critical values found analytically using an aspect ratio of A=10, i.e. the critical value of the Rayleigh number we found analytically is in good agreement with that found by Vasseur. The difference between the results obtained numerically and those obtained analytically by Galerkin's method may be due either to the assumptions made in determining the analytical solution, or to the aspect ratio used in the numerical simulation, since linear stability was studied for an infinite cell. The linear stability study has enabled us to define the range of Ra values for which the single-cell flow remains stable, thus enabling separation of the mixture constituents.

8. Conclusion:

In this chapter, we have explored the study of a planar layer subjected to horizontal heat flow, highlighting the underlying mechanisms of thermogravitation and their impact on the separation of the constituents of a binary mixture. Through an analytical and numerical approach, we have demonstrated that the horizontal configuration presents distinct characteristics that enable efficient separation of species under the effect of thermal gradients.

Analysis of the equations governing heat and mass transfer in this system revealed that the temperature gradient imposes a flow of material, which in turn is influenced by convection. By considering the appropriate boundary conditions and using suitable solving methods, we were able to establish analytical expressions for the temperature and concentration profiles, enabling us to assess the dynamic behavior of the mixture. These results were compared with numerical data obtained using the finite element method, showing satisfactory agreement with the results of Vasseur et al [33], which validates our theoretical approach.

The results obtained show that, unlike vertical configurations, the horizontal configuration enables significant separations to be achieved with greater layer thicknesses. We have identified an optimal layer thickness that maximizes separation while minimizing the relaxation time required to reach stationary equilibrium. This finding underlines the importance of experimental design in separation systems, where larger dimensions can facilitate the practical implementation of thermogravitation devices.

We have also demonstrated the impact of Rayleigh number on separation efficiency. At low Rayleigh numbers, thermodiffusion effects predominate, leading to higher separations. On the other hand, at high values, intense convection can homogenize the mixture, reducing separation efficiency. This highlights the need for optimal coupling between convection and thermodiffusion to achieve maximum results.

However, certain limitations need to be considered. Although this study has provided a solid foundation for understanding thermogravitational separation in horizontal configurations, edge effects, density variations and the non-linear nature of the equations need to be taken into account in future studies. Further experimental work is needed to validate the theoretical models and explore more complex mixtures, as well as varied operating conditions.

In conclusion, this research into thermogravitation in a flat layer subjected to horizontal heat flow has not only broadened our understanding of separation mechanisms, but has also paved the way for promising industrial applications. The results obtained provide valuable pointers for the development of new fluid separation technologies, contributing to advances in the fields of energy, the environment and chemical processes. Future prospects include exploring hybrid configurations and optimizing existing systems for enhanced performance in real-life applications.

Conclusion

The study carried out on thermogravitational separation in porous media has demonstrated the importance of coupling natural convection and thermodiffusion to improve the separation of components in a binary mixture. By exploiting innovative configurations, such as a horizontal cavity filled with porous media and subjected to a constant vertical heat flow, this research has contributed to the optimization of separation processes in complex systems.

The phenomenon of thermogravitation, resulting from the interaction between a temperature gradient and a gravitational field, enhances the effects of thermodiffusion. In the presence of porous media, this phenomenon is further amplified by flow modulation and enhanced concentration gradients. The approach adopted in this work relies on rigorous analytical modelling, based on Darcy's equations, and numerical simulations using the finite element method, enabling us to accurately predict the optimum conditions for obtaining maximum separations.

The results obtained show that the horizontal configuration studied offers significant advantages over conventional vertical devices, particularly in terms of ease of experimental set-up and control of geometric parameters. In addition, the use of a porous medium enables convective flows to be better controlled, and higher separations to be achieved, even for complex mixtures. These results are in line with previous work on thermogravitation and pave the way for new applications, notably in the fields of geochemistry, industrial fluid processing and isotope separation.

However, despite these advances, a number of challenges remain. Experimental validation of numerical and analytical results under real-life conditions is essential to confirm theoretical predictions. What's more, extending these models to more complex multicomponent mixtures and varied geometric configurations could open up new prospects for the development of industrial processes based on thermogravitational separation.

In conclusion, this study has contributed to a better understanding of thermogravitational separation mechanisms in porous media, and highlighted innovative configurations for optimizing these processes. The results provide a solid basis for the development of new fluid separation technologies, with potential applications in several areas of fluid and energy engineering. Future work could focus on integrating these methods into large-scale industrial processes, as well as improving numerical modeling for more complex systems.

Appendix 1

MAPLE software solves the problem of thermogravitational diffusion in a horizontal cell with constant flux applied to the horizontal walls.

```
> restart;
> eqFc:=diff(Fc(y),y$2)=-Ra*(b+phi*m):
> soleqFc:=dsolve(eqFc,Fc(y)):
> assign(soleqFc):
> clFc1:=subs(y=0,Fc(y))=0:
> clFc2:=subs(y=1,Fc(y))=0:
> solclFc:=solve({clFc1,clFc2},{_C1,_C2}):
> assign(solclFc):
> Fc(y):=factor(simplify(eval(Fc(y)))):
On pose : PSI0=(Ra*(b+phi*m))/2
> Fc(y):=PSI0*(y-y^2):
> u(y):=factor(simplify(diff(Fc(y),y))):
> eqtemp:=diff(f(y),y$2)=u(y)*b:
> soleqtemp:=dsolve(eqtemp,f(y)):
> assign(soleqtemp):
> cltemp1:=subs(y=0,diff(f(y),y))=-1:
> sol1temp:=solve({cltemp1},{_C3}):
> assign(sol1temp):
> f(y):=simplify(factor(eval(f(y)))):
> cltemp3:=int(u(y)*f(y)-b,y=0..1):
> f(y):=factor(eval(simplify(f(y)))):
> _C4:=0:
> T:=eval(b*x+f(y)):
> eqconc:=u(y)*m=(1/Le)*(diff(g(y),y$2)-diff(f(y),y$2)):
> soleqconc:=dsolve(eqconc,g(y)):
> assign(soleqconc):
> clconc1:=subs(y=0,diff(g(y),y)-diff(f(y),y))=0:
> sol1eqconc:=solve({clconc1},{_C5}):
> assign(sol1eqconc):
> g(y):=factor(simplify(eval(g(y)))):
> c:=m*x+g(y):
> clnonpen:=factor(simplify(int(Le*u(y)*g(y)-m+b,y=0..1)=0)):
> clconserv:=simplify(int(int(c,x=0..A),y=0..1)=0):
> resuc4:=solve({clconserv},{_C6}):
> assign(resuc4):
> c:=factor(simplify(eval(c))):
> soleq2:=solve({clnonpen},{m}):
> assign(soleq2):
> soleq1:=solve({cltemp3},{b}):
> assign(soleq1):

                                                        >
```

```
eqpsi0:=simplify(PSI0=(Ra/2)*(b+phi*m)):
>eq:=collect(factor(simplify(PSI0*2*(PSI0^2+30)*(Le^2*PSI0^2+30)-
5*Ra*PSI0*(Le^2*PSI0^2+30+30*phi*Le+30*phi))),PSI0):
> soleq:={solve({eqpsi0},{PSI0})}:
> factor(simplify((Ra*Le^2-12-12*Le^2)^2-(Ra^2*Le^4+24*Ra*Le^2-24*Ra*Le^4+144-
288*Le^2+144*Le^4+48*Le^3*Ra*phi+48*Le^2*Ra*phi))):
> plt1:=plot([S],Ra=0..0,thickness=[1],color=[black]):
>plt2:=PLOT(POINTS([1.5,4.248875],[3,4.223495],[2,4.6257725],[4,3.6816775],[5,3.192
6225],[6,2.7610025],[7,2.37237],[8,2.0116375],[9,1.6642075],[10,1.31356],[11,0.9368525
],[12,0.5269825],[13,0.2627875],[14,0.15694],[15,0.1037575],[18,0.03599],[22,0.005365],
[26,0.00707],[28,0.01058],[30,0.0130875],[36,0.0172225],[40,0.018465],[44,0.019095],[5
0,0.0194075],[60,0.0191375],[70,0.018505],[90,0.0170725],[100,0.016405],[150,0.01384],
[250,0.1037575],[350,0.0095025],[700,0.0069475],[1000,0.00586],[1500,0.0047175],SYM
BOL(DIAMOND)))):
> display({plt1,plt2}):
```

Appendix 2

Approaches to studying linear stability

```
> restart;
> N:=5:Le:=2:phi:=2/5:eps:=1:sigma:=I*omega:Digits:=30:
> Fc:=sum(a[i]*sin(Pi*i*y),i=1..N):
> theta:=(sum(b[i]*cos(Pi*i*y),i=0..N-1)):
> sep:=(sum(c[i]*cos(Pi*i*y),i=0..N-1)):
> Fc0(y):=PSI0*(y-y^2):
> T0:= b*x+1/2*PSI0*b*y^2-1/3*PSI0*b*y^3-y:
> c0:=m*x+1/2*PSI0*m*Le*y^2-1/3*PSI0*m*Le*y^3+1/2*PSI0*b*y^2-
1/3*PSI0*b*y^3-y-1/12*PSI0*b-1/12*PSI0*m*Le+1/2-1/2*m*A:
> expand(subs(y=1,diff(sep,y))):evalf(subs(y=0,diff(sep,y))):
> expand(subs(y=1,diff(theta,y))):evalf(subs(y=0,diff(theta,y))):
Calcul des résidus :
> loperFc:=diff(Fc,y,y)-k^2*Fc:loperT:=diff(theta,y,y)-k^2*theta:lopersep:=diff(sep,y,y)-
k^2*sep:loperFc2:=diff(loperFc,y,y)-k^2*loperFc:
> Eq1:=loperFc+I*k*Ra*(phi*sep+theta):
> Eq2:=loperT-sigma*theta-b*diff(Fc,y)+Fc*I*k*diff(T0,y)-diff(Fc0(y),y)*I*k*theta:
> Eq3:=(lopersep-loperT)/Le-eps*sigma*sep-diff(Fc,y)*m+I*k*Fc*diff(c0,y)-
diff(Fc0(y),y)*I*k*sep:
```

Linear stability program

Projection

```
> with(LinearAlgebra):
> sys:=seq(int(Eq1*sin(Pi*i*y),y=0..1),i=1..N),seq(int(Eq2*cos(Pi*i*y),y=0..1),i=0..N-
1),seq(int(Eq3*cos(Pi*i*y),y=0..1),i=0..N-1):
> (aa,bb):=GenerateMatrix([sys],[seq(a[i],i=1..N),seq(b[i],i=0..N-1),seq(c[i],i=0..N-1)]):
> deta1:=Determinant(aa):
> eqc:=evalc(deta1):
>eqi:=115750215/128*Pi^22*k^2*omega^3+53947935/2048*Pi^14*k^2*omega^7-
2071633815/4096*Pi^18*k^2*omega^5-
626101857/1024*Pi^16*k^4*omega^5+14235/32768*Pi^2*k^14*omega^7-
16384/2734375*m^4*k^16*Ra^4*omega-............
> eqr:=eqc-I*eqi:
> PSI0:=1/2*sqrt(5)*sqrt(Ra*Le^2-12-12*Le^2+sqrt(Ra^2*Le^4+24*Ra*Le^2-
24*Ra*Le^4+144-288*Le^2+144*Le^4+48*Le^3*Ra*phi+48*Le^2*Ra*phi))/Le:
> b:=(5*PSI0)/(PSI0^2+30):
> m:=-(Le*PSI0^2*b-5*Le*PSI0-30*b)/(Le^2*PSI0^2+30):
> eqrv:=eval(eqr):
> eqiv:=eval(eqi):
> with(plots):
```

```
>implicitplot3d([eqrv,eqiv],k=1..10,omega=1..200,Ra=1..500,color=[red,blue],numpoints=
3000):
> for k1 from 4.3 to 4.82 by 0.01 do
solfin1:=fsolve({subs(k=k1,eqrv)=0,subs(k=k1,eqi)=0},{Ra,omega},{Ra=250..320,omega
=80..200}):lprint(`k= `,k1,`sont : `,evalf(solfin1,10)):od:
```

References

[1] BIERLEIN J., A., A Phenomenological Theory of the Soret Diffusion, Journal of Chemical Physics, 23, pp. 10-14 (1955).

[2] CASIMIR H. G., Statistical Foundations of the Thermodynamic of Irreversible Processes, Chem. Weekblad 50, pp. 318-320 (1954).

[3] CLUSIUS K. and DICKEL G., Neues Verfahren zur Gasentmishung und Isopentrenung, Naturwiss. 26, p. 546 (1938).

[4] COSTESEQUE P., Sur la Migration Sélective des Isotopes et des Elements par la Thermodiffusion dans les solutions. Application de l'Effet Thermogravitationnel en Milieux Poreux. Observation expérimentale et Conséquences Géochimiques, Doctoral thesis, Paul Sabatier University, Toulouse (1982).

[5] COSTESEQUE P., HRIDABBA M. and SAHORES J., Possibilité de Différenciation des Hydrocarbures par Diffusion Thermogravitationnelle dans un pétrole Brut Imprégnant un Milieux Poreux, C. R. Acad. Sci. 304, 17, p. 1069-1074 (1987).

[6] COSTESEQUE P., POLLAK T., PLATTEN J. K., and MARCOUX M., Transient-State Method for Coupled Evaluation of Sor Thermodynamics, (New York: Dover 1984).

[7] DE GROOT S. R. and MAZUR P., Non-Equilibrium Thermodynamics (New York: Dover 1984).

[8] DE GROOT S. R. , L'Effet Soret, Diffusion Thermique dans les Phases Condensées, Doctoral thesis, University of Amsterdam (1945).

[9] DE GROOT S and MAZUR P., Non Equilibrium Thermodynamics, North-Holland pub.Co (1961).

[10] ECENARRO O., MADARIAGA A., NAVARRO J., SANTAMARIA C. M., CARRION J. A., and SAVIRON J. M., Non Steady State Density Effects in Liquid Thermal Diffusion Columns, J . Phy. Condens. Matter 1, p. 9741-9750 (1989).

[11] Elhajjar B., Charrier-Mojtabi MC., Mojtabi A., Separation of a binary fluid mixture in a porous horizontal cavity, Phys Review E, Vol: 77, Article Number: 026310 (2008)

[12] Elhajjar B., Mojtabi A., Marcoux M., et al, Study of thermogravitation in a horizontal fluid layer, COMPTES RENDUS MECANIQUE vol: **334**. p. 621-627 (2006)

[13] El Maataoui M., Cons équence de la Thérmodiffusion en Milieu Poreux sur l'Hydrolyse de Solutions de Chlorures Ferriques et sue les Migrations d'Hydrocarbures dans les Mélanges de N-Alcanes et dans un Pétrole Brut : Implications Géochimiques, Thèse de Doctorat 3ᵉᵐᵉ cycle, Université Paul Sabatier, Toulouse (1986).

[14] EMERY A. H. and LORENZ M., Thermal diffusion in packed column; A. I. Ch. E. J. 9, pp. 661-663 (1963).

[15] ESTEBE J., Etude de l'Effet Thermogravitationnel en Milieux Poreux, Thèse Docteur Ingénieur, n° 240, Université Paul Sabatier, Toulouse (1970).

[16] FURRY W. H., JONES R. C. and ONSAGER L., On the Theory of Isotropic Separation by Thermal Diffusion, Physical Review, 55, pp. 1083-1095 (1939).

[17] KÖHLER W. and WIEGAND S. 2002: Thermal Nonequilibrium Phenomena in Fluid Mixtures (Springer Lecture Notes in Physics vol.584) (Berlin: Springer).

[18] KORSGHING W. J. and WIRTZ K., Separation of Liquid Mixtures in the Clusius Separation Tube (Separation of Zinc Isotopes),Naturwissenschaften 27, pp. 367-368 (1939).

[19] LORENZ M. and EMERY A. H., The Packed Thermodiffusion Column, Chemical Engineering Science, 11, pp. 16-23 (1959).

[20] LUDWIG C. D., Diffusion Zwischen Ungleigherewärmten Oretn Gleich Zusammengesetzter. Lösunger, Akad. Wiss. Wien, Math. Nturw. Kl. 20, p. 539 (1856).

[21] MAJUNDAR S. D., Theory Separation Isotopes by Thermal Diffusion, Physical Review, 81 (5), p. 844 (1951).

[22] MARCOUX M., Contribution à l'Etude de la Diffusion Thermogravitationnelle en Milieux Poreux, PhD thesis, Université Paul Sabatier, Toulouse (1998).

[23] ONSAGER L., Reciprocal Relations in Irreversible Process, I. Phys. Rev. 38, pp. 405-426 (1931).

[24] PLATTEN J. K., BOU-ALI M. M., and DUTRIEUX J. F., Enhanced Molecular Separation in Inclined Thermogravitational Columns, J. Phys. Chem. B, vol. 107, p. 11763-11767 (2003).

[25] PLATTEN J. K., and LEGROS J. C.,Convection in Liquids , Springer Verlag (1984).

[26] PRIGOGINE I., Thermodynamics of Irreversible Processes, I. Phys. Rev. 38, pp. 405-426 (1961).

[27] RIVIERE E. Sur la Migration des Composés Hydrocarburés dans des Mélanges Complexes (Huiles Naturelles) par Diffusion Thermogravitationnelle en Milieu Poreux. Etude Expérimentale et Etat des Modélisations, PhD thesis, Paul Sabatier University, Toulouse (1991).

[28] SANCHEZ V., La Diffusion Thermique et son Application au Fractionnement des Mélange Binaires, PhD thesis, Université Paul Sabatier (1975).

[29] SCHOTT J., Contribution à l'Etude de la Thermodiffusion dans les Milieux Poreux. Application aux possibilités de Concentrations Naturelles, PhD thesis, Université Paul Sabatier, Toulouse (1973).

[30] SORET C., Influence de la Température sur la Distribution des Sels dans leurs Solutions, C. R. Acad. Sci. 91, p.289-291 (1880).

[31] SORET C., Sur l'Etat d'Equilibre que Prend , du Point de Vue de sa Concentration, une Dissolution Saline Primitivement Homogéne dont deux Parties sont Portées à des Températures Différentes, Ann. Chim. Phys. 22, p.293-297 (1881).

[32] SULLIVAN L. J., RUPPEL T. C. and WILLINGHAM C. B.,Packed Thermal diffusion columns, Ind. Eng. Chem. 49, pp. 110-113 (1957).

[33] VASSEUR P., BAHLOUL A., BOUTANA N., Double Diffusive and Soret-Induced Convection in a shallow Horizontal Porous layer, Journal of Fluid Mechanics, vol. 491. p. 315-352 (2003).

I want morebooks!

Buy your books fast and straightforward online - at one of world's fastest growing online book stores! Environmentally sound due to Print-on-Demand technologies.

Buy your books online at
www.morebooks.shop

Kaufen Sie Ihre Bücher schnell und unkompliziert online – auf einer der am schnellsten wachsenden Buchhandelsplattformen weltweit! Dank Print-On-Demand umwelt- und ressourcenschonend produzi ert.

Bücher schneller online kaufen
www.morebooks.shop

info@omniscriptum.com
www.omniscriptum.com